Bibliographic information published by the German National Library:

The German National Library lists this publication in the National Bibliography; detailed bibliographic data are available on the Internet at http://dnb.dnb.de .

Imprint:

Print and binding: Books on Demand GmbH, Norderstedt Germany
ISBN: 9783668566651

This book at GRIN:

http://www.grin.com/en/e-book/379719/performance-indices-of-a-power-plant-using-exergy-based-analyses

Zin Eddine Dadach

Performance indices of a power plant using exergy-based analyses

GRIN Publishing

ASSESSMENT OF PERFORMANCE INDICES OF A POWER PLANT USING EXERGY-BASED ANALYSES

ABSTRACT

The objective of this applied industrial research was to conduct an exergy-based analysis for an Open Cycle Gas Turbine (OCGT) in Abu Dhabi (UAE) in order to evaluate its performance under design conditions and during summer weather conditions. The first explanation for this investigation is that CO_2 emissions from power generation plants in the UAE are responsible for about 33% of the 200 Million tons of the total CO_2 emitted in 2013 in the country [1]. The second reason for this industrial project is that the standard conditions used for the design of gas turbines are 288K, sea level atmospheric pressure and 60% relative humidity [2]. However, the average summer weather conditions in Abu Dhabi are T=316K and a relative humidity of 50%. As a consequence, the effects of summer weather conditions on different performance indices of the power plant were also studied. Aspen Hysys V8.6 with the Soave-Redlich-Kwong (SRK) equation were used as tools to simulate the power plant under standard conditions

For the first performance index under investigation Y_D (exergy destruction ratio), results of the exergy analysis showed that the combustor is the main contributor to the exergy destruction of the power plant. Summer conditions increased its exergy destruction ratio by 21.2 % and decreased its exergetic efficiency by 2.6%. However, due to the positive effects of higher temperatures, the summer weather conditions decreased the exergy destruction ratio of the turbine by 55% and increased its exergetic efficiency by 4.3%.

The contribution of the cost of the exergy destruction in each equipment of the plant to the total cost of the final product (electricity) of the power plant was investigated using an exergoeconomic analysis of the plant and the second performance index r_k (relative cost difference). In accordance with the first investigation, the combustor had also the highest contribution to the cost of the final product (r=15.4%). The summer weather conditions had increased the cost rate of its exergy destruction by 19.8%. and its relative cost difference by 18.2 %. On the other hand, the summer weather conditions have decreased the cost of exergy destruction of the turbine by 14.3 %. and its contribution to the cost of the final product (r) by 31.6%.

The effects of summer weather conditions on the environmental impact of the power plant were investigated using two different performance indices. The relative difference of exergy-related environmental impacts (r_b) was utilized for each equipment of the power plant and the environmental impact of a kWh of electricity (EIE) was used for the power plant. In agreement to the exergetic analysis, the results of the exergoenviromental analysis indicated that the combustor presents the highest environmental impact of exergy destruction. The summer weather conditions further increased this impact by 21.6%. In addition, the combustor also had the highest contribution to the total environmental impact of the final product (rb =14.4%), while summer weather conditions increased this contribution by 20.8%. The expander had the second highest environmental impact of exergy destruction and summer weather conditions decreased this impact by 53.7%. The expander had the lowest contribution to the total environmental impact of the final product (rb =8.4%), while summer weather conditions decreased this contribution by 58.3%.

The environmental impact of a kWh of electricity of the power plant under standard conditions was 37.8 mPts/kWh (exergy destruction only), and 54.7 mPts/kWh (both exergy destruction and exergy loss). The summer weather conditions increased these impacts by 6.6% and 10.7 % respectively. It could be concluded that the negative effects of summer weather conditions on the performance of the combustion chamber and the compressor were partly compensated by their positive effects on the performance of the turbine.

RESEARCH TEAM

EXERGY ANALYSIS

Omar Mohamed Alhosani[1], Abdulla Ali Alhosani[1]

EXERGOECONOMIC ANALYSIS

Ahmed Nabil Al Ansi[1], Mubarak Salem Ballaith[1], Hassan Ali Al Kaabi[1]

EXERGOENVIRONMENTAL ANALYSIS

Mohamed Mohamed Alhammadi [1], Mubarak Haji Alblooshi[1], Fontina Petrakopoulou [2]

PRINCIPAL INVESTIGATOR (PI):

Zin Eddine Dadach[1]

[1]Department of Chemical and Petroleum Engineering, Higher Colleges of Technology, Abu Dhabi, UAE

[2] Department of Thermal and Fluid Engineering, University Carlos III of Madrid, 28911 Madrid, Spain

NOMENCLATURE

AC amortization cost

$\hat{C}$ annualized cost

ACF actual cost factor

B_j environmental impact rate of the j-th material stream (Eco-indicator 99) (mPts/s)

b_j specific environmental impact rate of the j-th material stream (Eco-indicator 99) (mPts/MJ)

CRF capital recovery factor

E exergy rate (MW)

ED exergy destruction (MW)

EE exergetic efficiency

EIE environmental Impact of Electricity produced (mPts/kWh)

EL exergy loss

e specific exergy (kJ/kg)

FEI fixed capital investment

f_b exergoenvironmental factor, which expresses the relative contribution of component-related environmental impact to the sum of environmental impacts associated with the component (%)

f_k exrgoeconomic factor, which express the relative contribution of component-related to the total capital investment (%)

h specific enthalpy (kJ/kg)

IC initial cost

HHV high Heating value (MJ/kg)

LHV low heating value (MJ/kg)

m mass flow rate (kg/s)

OCGT Open Cycle Gas Turbine

P Pressure

PEC purchased equipment cost

Q heat rate (MW)

r_k relative cost difference

r_b relative difference of exergy-related environmental impacts (dimensionless)

s specific entropy (kJ/kg.K)

T temperature

W work rate (MW)

Y component-related environmental impact rate associated with the life cycle of the component (Eco-indicator 99) (mPts/s)

y exergy destruction ratio, which compares the exergy destruction within component with the exergy destruction within the overall system (%)

$\hat{Z}$ capital cost rate

Subscripts

CC combustor

Ch chemical

CV control volume

D destruction

F fuel

fg fuel gas

i chemical species

j j-th stream

K compressor

k k-th component of the plant

L lost

P product

Ph chemical

Q heat

T total

TB turbine

W work

0 dead state

Superscripts

i chemical species

PF pollutants formation

INTRODUCTION

The inefficiencies linked to irreversibilities in energy-conversion systems are invisible in a typical energy balance (1st Law of Thermodynamics) but can be evaluated using an exergy analysis (2nd Law of Thermodynamics). According to the theory, the exergy destroyed in each component of a power generation plant represents the loss of performance related to irreversibilities. To enhance the performance of the plant, efforts will be mainly focused on the equipment that presents the highest exergy destruction since it will offer the largest improvement of the exergy efficiency of the plant. The first performance index under investigation is the exergy destruction ratio Y_D. In order to investigate the summer weather conditions on this performance index, an exergy analysis of the plant will be conducted during a typical summer day and compared to the results under standard design conditions

The final objective of increasing exergy efficiency of power plants is to reduce the consumption of fuel and minimize their environmental impact. In exergoeconomic and exergoenvironmental analyses, exergy destruction costs and environmental impacts are linked to irreversibilities [3-4]. The combination of exergy with costs was first used by Keenan [5] and the concept of exergoeconomics was introduced by Tsatsaronis [6].The exergoeconomic accounting method is used in the design and operation of gas turbines to calculate the costs of final products as well as the costs of the exergy destroyed within each equipment. It can be considered as an exergy-aided cost reduction approach that uses the exergy costing principle [7]. Following the exergy analysis of the power plant, a detailed exergoeconomic analysis of the plant based on Specific Exergy Costing (SPECO) method is presented in this investigation. One of the objectives of the analysis is to compare the values of the performance index r_k (contribution to the cost of the final product of the power plant) calculated during summer atmospheric conditions to the values obtained under design conditions.

In order to ensure sustainable operation, the environmental impact of the plant must also be investigated. To achieve this goal, the environmental impact of each plant's component must be compared to its corresponding impact under standard conditions. The final step of this investigation is to study the environmental impact of the plant by comparing the values of the performance index r_b (environmental impact difference) of each component of the plant during a typical summer day and under standard conditions. The effects of summer weather conditions on the power plant will be investigated by comparing the values of the performance index EIE (environmental impact of electricity) during summer weather conditions and under standard conditions.

BACKGROUND

Concept of Exergy

Exergy is commonly defined as the maximum theoretical work that can be extracted from a "combined system" consisting of a "system" under study and its "environment" as the system passes from an initial state to a state of equilibrium with the environment [8]. When a system is in equilibrium with the environment, the state of the system is called 'dead state' and its exergetic value is zero. According to Bejan et al.[7], total exergy (E_T) of a stream is constituted by four main components:

$$E_T = E_{ph} + E_{Ch} + E_k + E_p$$

(1)

The physical exergy (E_{ph}) is often described as the maximum theoretical useful work obtainable as the system passes from its initial state (P, T) to the "restricted dead state" (P_0, T_0). On the other hand, the chemical exergy (E_{ch}) is the maximum useful work obtainable as the system passes from the "restricted dead state", where only the conditions of mechanical and thermal equilibrium are satisfied, to the "dead state" where it is in complete equilibrium with the environment [9]. The kinetic (E_k) and potential (E_p) exergies are associated to the system velocity and height, respectively measured relative to a given reference point. When a system is at rest relatively to the environment (E_k=E_p=0), the total mass specific exergy (e_T) of a stream is defined as:

$$e_T = e_{ph} + e_{Ch}$$

(2)

Standard Chemical Exergy of a gas mixture

The chemical exergy per mole of gas (k) is given by the following equation [9]:

$$e_{Ch}^{k} = -R.T.\ln x_e^k$$

(3)

For a mixture of gases, the chemical exergy per mole of the mixture could be estimated using [9]:

$$e_{Ch} = \sum x_k . e_{Ch}^k + R.T.\sum x_k . \ln x_k$$

(4)

The exergy of fuel is equivalent to the calculated reversible work. The values of exergy of hydrocarbons and other components are listed in the literature [3] and the chemical exergy of a fuel could be estimated using equation (4). It should be noted that the value of the specific chemical exergy of a fuel at dead-state conditions is between the lower (LHV) and higher (HHV) heating values of the fuel [9].

Exergy Balance in Open Systems

Unlike energy, exergy is not conserved in any real process. As a consequence, an exergy balance must contain a "destruction" term, which vanishes only for a reversible process. The general form of exergy balance for a control volume could be written as [9]:

$$\frac{dE_{CV}}{dt} = \Sigma E_{heat} + E_{work} + \Sigma m_i \, . e_{T,i} - \Sigma m_e . e_{T,e} - E_D$$

(5)

For a steady state system, equation (5) could be rewritten as:

$$0 = \Sigma E_{heat} - W_{cv} + \Sigma m_i \, . e_{T,i} - \Sigma m_e . e_{T,e} - E_D$$

(6)

In equation (6), the total specific exergy transfer at the inlets and outlets could be written as:

$$e_T = (h - h_0) - T_0 \, (s - s_0) + \sum x_k \, . e_{Ch}^k + R.T. \sum x_k \, . \ln x_k$$

(7)

h and s are properties at the inlet and the outlet of the system. h_0 and s_0 are respectively the specific enthalpy and the specific entropy of the restricted dead state.

Exergy analysis of an Open Cycle Gas Turbine

As shown in Figure 1, the power plant under investigation is an Open Cycle Gas Turbine (OCGT). Exergy destruction (ED) within a component of a power generation plant is the measure of irreversibility that is the source of performance loss. An exergy analysis will determine the magnitude and the source of thermodynamic inefficiencies in a power plant.

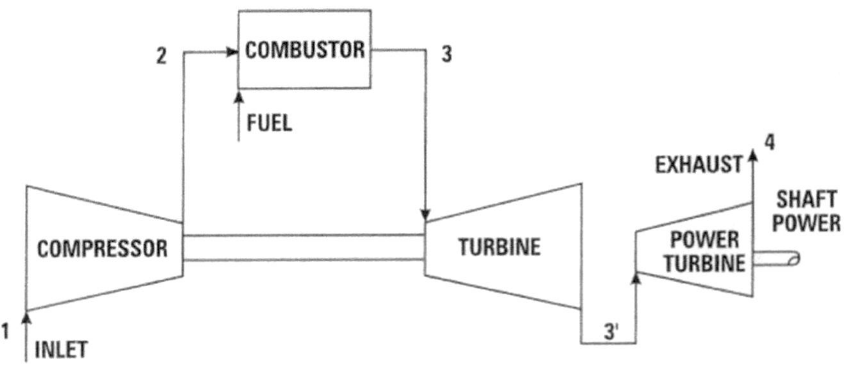

Figure 1: Schematic Diagram of an Open Cycle Gas Turbine [7]

Based on Figure 1, the exergy destruction (ED) and exergy efficiency (EE) for the three main component of an open cycle gas turbine (OCGT) are defined using the equations [7]:

Compressor

$$(\mathrm{ED})_{\mathrm{K}} = \mathrm{W_K} - \ \mathrm{m_{air}}\ (\mathrm{ex_{T2}} - \mathrm{ex_{T1}}) \tag{8}$$

$$(\mathrm{EE})_{\mathrm{K}} = 1 - \frac{(\mathrm{ED})_{\mathrm{K}}}{\mathrm{W_K}} \tag{9}$$

Combustor

$$(\mathrm{ED})_{\mathrm{cc}} \ = \ \mathrm{m_{air}}\mathrm{x}\ \mathrm{ex_{T2}} + \mathrm{m_{fuel}}\mathrm{x}\ \mathrm{ex_{fuel}} - \mathrm{m_{fg}}.\,\mathrm{ex_{T3}} \tag{10}$$

$$(\mathrm{EE})_{\mathrm{cc}} = \ 1 - \frac{(\mathrm{ED})_{\mathrm{cc}}}{\mathrm{m_{air}}\mathrm{x}\ \mathrm{ex_{T2}} + \mathrm{m_{fuel}}\mathrm{x}\ \mathrm{ex_{fuel}}} \tag{11}$$

Turbine

$$(\mathrm{ED})_\mathrm{T} = \mathrm{m_{fg}x}\,(\mathrm{ex_{T3}} - \mathrm{ex_{T4}}) - \mathrm{W_T}$$

(12)

$$(\mathrm{EE})_\mathrm{T} = 1 - \frac{(\mathrm{ED})_\mathrm{T}}{\mathrm{m_{fg}x}\,(\mathrm{ex_{T3}} - \mathrm{ex_{T4}})}$$

(13)

For a power generation plant producing electricity using the combustion of fuel, the rate of product exergy of the kth equipment ($\hat{\mathrm{E}}_{\mathrm{P,k}}$) is the exergy of the desired output resulting from the operation of the component, while the rate of fuel exergy of the same component ($\hat{\mathrm{E}}_{\mathrm{F,k}}$) is the expense in exergetic resources for the generation of the desired output. The rate of fuel exergy and product exergy of the three main components are defined in Table1.

Table 1: Rate of fuel and product exergy for each component [7]

Equipment	$\hat{\mathbf{E}}_{\mathbf{F,k}}$ (MW)	$\hat{\mathbf{E}}_{\mathbf{P,k}}$ $(\boldsymbol{MW})$
Compressor	W_K	$m_{air}\,(ex_{T2} - ex_{T1})$
Combustor	$m_{air}x\;ex_{T2} + m_{fuel}x\;ex_{fuel}$	$m_{fg}.\,ex_{T3}$
Turbine	$m_{fg}x\,(ex_{T3} - ex_{T4})$	$W_P + W_K$

The rate of exergy destruction within the kth component, $(\mathrm{ED})_\mathrm{k}$, is calculated as the difference between its rate of fuel and product exergy [7]:

$$(ED)_k = \hat{E}_{F,k} - \hat{E}_{P,k}$$

(14)

And the exergy destruction ratio in each equipment could be written as:

$$Y_{D,k} = \frac{(ED)_k}{\hat{E}_{F,T}}$$

(15)

Economic analysis

The economic model takes into account the cost of the components, including amortization and maintenance, and the cost of fuel consumption. The first stage of an economic analysis of a power plant is to estimate the *purchased equipment cost* (PEC). The capital needed to purchase and install equipment is called the fixed capital investment (FCI). A number of methods have been published to estimate the purchase cost of equipment based on design parameters [7, 10-11]. The levelized cost method of Moran [12] is considered in this investigation. The amortization cost (AC) for a component (k) of the power plant depends on the initial cost (IC) and the actual cost factor (ACF):

$$(AC)_k = (IC)_k - [S_n x\ ACF\ (i,n)]_k$$

(16)

The annualized cost of an equipment (k) of the power plant could be estimated using the capital recovery factor (CRF):

$$(\hat{C})_k\ (\$.year^{-1}) = (AC)_k x\ CRF(\ i,n)_k$$

(17)

The capital recovery factor (CRF) is a function of the interest (i) and the estimated equipment life (n) [11]:

$$CRF = \frac{ix(1+1)^n}{(1+i)^n - 1}$$

(18)

CRF's value of 0.131 is calculated based on an interest rate of 10% and a total operating period of 15 years. The capital cost rate for the equipment (k) of the plant [7]:

$$\hat{Z}_K = \frac{\hat{C}_k x\ \phi_k}{3600x\ N}$$

(19)

$\hat{C}_k$, ϕ_k and N are respectively the annualized cost of the equipment, the maintenance factor and the annual number of operating hours. Considering typical values of the maintenance factor (ϕ_k =1.06) and the annual number of operating hours (N=8000)

The cost function of the three main equipment of an open cycle gas turbine could be written as [7]:

Compressor

$$\hat{Z}_K = 71.1x\ m_1 x \frac{1}{(0.9 - \eta_K)} x P_r x ln(P_r) x 4.84\ x\ 10^{-9}$$

(20)

Combustor

$$\hat{Z}_{cc} = \left(46.08xm_1 x\left(1 + e^{(0.018xT_3 - 26.4)}\right)x(0.995 - (P_3/P_2))^{-1} x\ 4.84\ x\ 10^{-9}\right.$$

(21)

Turbine

$$\hat{Z}_T = \left[479.34 x m_{fg} x \,(0.92 - \eta_{sT})^{-1} x ln \left(\frac{P_3}{P_4}\right) x \left(1 + e^{(0.036 x T_3 - 54.4)}\right)\right] x 4.84 \, x \, 10^{-9}$$
(22)

Exergoeconomic analysis

Exergy costing of a power generation plant involves cost balance for each component separately. In a cost balance around the k-th component, the total cost of all exiting streams (e) is equal to the total cost of the entering streams (i) plus the appropriate charges due to capital investment and the expenses for operations and maintenance ($\hat{Z}_k$). Table 2 represents the cost balances and the auxiliary equations of the three main equipment of the power plant:

Table 2: Cost balances and auxiliary equations [5-7]

Equipment	**Cost balance**	**Auxiliary equations**
Compressor	$\hat{C}_2 = \hat{C}_1 + \hat{C}_{W,K} + \hat{Z}_K$ (23)	$\hat{C}_1$ =0 (24)
Combustor	$\hat{C}_3 = \hat{C}_2 + \hat{C}_F + \hat{Z}_{CC}$ (25)	
Turbine	$\hat{C}_4 + \hat{C}_{W,k} + \hat{C}_P = \hat{C}_3 + \hat{Z}_T$ (26)	$\frac{\hat{C}_{W,k}}{W_K} = \frac{\hat{C}_{W,T}}{W_T}$ (27) and $\frac{\hat{C}_4}{E_4} = \frac{\hat{C}_3}{E_3}$ (28)

Cost of exergy destruction

Invisible in cost balance equations, the exergy destruction cost for each equipment of the power plant could be estimated by combining the equations related respectively to the exergy destruction (ED) and the cost rate balance. Two approaches are usually used to approximate the average cost associated to the exergy destruction within the equipment (k) of the plant. In the first approach, the product exergy rate ($\hat{E}_{P.k}$) is assumed fixed and the unit cost of fuel of the k component ($c_{F,k}$) is independent of the exergy destruction. In this situation, the cost of exergy destruction represents the cost rate of additional fuel that must be supplied to the component (k) of the power plant to compensate the exergy destruction within the equipment and is defined as [7]:

$$\hat{C}_{D,k} = c_{F,k} \, x \, (ED)_k$$

(29)

In the second approach, the fuel exergy rate ($\hat{E}_{F.k}$) is considered constant and the unit cost of product of the (k) component ($c_{P,k}$) is independent of the exergy destruction. In this case, the cost of exergy destruction represents the loss of product due to the exergy destruction within the equipment (k) and is defined as [7]:

$$\hat{C}_{D,k} = c_{P,k}\ x\ (ED)_k$$

(30) The first approach is utilized in this investigation. The average unit cost of fuel ($c_{F\,k}$) and product ($c_{P,k}$), the cost rate of exergy destruction ($\dot{C}_{D,k}$), relative cost difference (r_k) and exergoeconomic factor (f_k) are very important factors for an exergoeconomic analysis of power generation plants [3]. The simplest way to estimate the cost of exergy destruction in a power generation plant is to consider the average cost per exergy unit of the fuel as the cost of exergy destruction unit. The cost rate of fuel and product for the three main components of the plant are defined in Table 3 [7].

Table 3: Cost rate of fuel and product for the main components of the power plant

Equipment	Cost rate of fuel $\hat{C}_F$ (\$/hr.)	Cost rate of product $\hat{C}_P$ (\$/hr.)
Compressor	$\hat{C}_{w,k}$	$\hat{C}_2 - \hat{C}_1$
Combustor	$\hat{C}_2 + \hat{C}_F$	$\hat{C}_3$
Turbine	$\hat{C}_3 - \hat{C}_4$	$\hat{C}_{W,T}$

The average cost of fuel ($c_{F,k}$), the average unit cost of product ($c_{P,k}$) could be estimated [10]:

$$c_{F,k} = \frac{\hat{C}_{F,k}}{\hat{E}_{F,k}} \qquad (31)$$

$$c_{P,k} = \frac{\hat{C}_{P,k}}{\hat{E}_{P,k}}$$

(32)

The relative increase in the average cost per exergy unit between the fuel and the product could be expressed by the relative cost difference (r_k) [7]:

$$r_k = \frac{c_{P,k} - c_{F,k}}{c_{F,k}} = \frac{1-EE_k}{EE_k} + \frac{\hat{Z}_k}{c_{F,k}\, x \hat{E}_{P,k}}$$

(33)

The exergoeconomic factor (f_k) is often used to evaluate the importance of the capital investment and operating and maintenance costs (non-exergy related costs) compared to the cost linked to the exergy destruction [10]:

$$f_k = \frac{\hat{Z}_k}{c_{F,k} x [(ED)_k] + \hat{Z}_k}$$

(34)

Exergoenvironmental analysis

An exergoenvironmental analysis reveals the relative importance of each plant component constituting an energy system, with respect to environmental impact. It also offers options for reducing the environmental impact of the plant. In an exergoenvironmental analysis, a one-dimensional characterization indicator is obtained using a Life Cycle Assessment (LCA). LCA is a technique used to quantify the environmental impact of inputs (resources) and outputs (products, pollutants, etc.) of systems relative to the natural use of resources, human health and other ecological areas. The quantification of the environmental impact caused by depletion and emissions of a natural resource used can be carried out using [13]:

(1). Life Cycle Assessment following ISO 14044 (2). Matrix-based LCA (3). Proxy measures

Proxy measures use a single value to represent the environmental impact of a product or material. An example of proxy measures is the life cycle impact assessment (LCIA) method Eco-indicator. The Eco-indicator of a material or a process is a number that indicates its environmental impact based on data from a life cycle assessment. The higher the indicator is, the greater the environmental impact of the process. LCIA methods, like Eco-indicator 95 [14], Eco-indicator 99 [15] and the Swiss Ecoscarcity [16] have been successfully utilized for energy conversion systems.

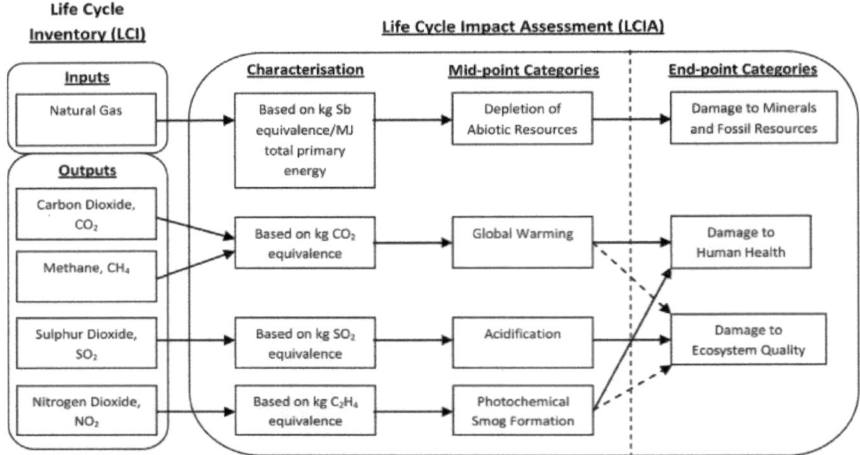

Figure 2: Typical LCA framework linking LCI via mid-point categories to end-point categories for selected damage types [17]

Eco-indicator 99 has been evaluated by various authors [18-21], with respect to its suitability in LCA-related issues and several LCA software packages support it (e.g., SimaPro and Gabi) [13]. As also seen in Figure 2, the Eco-indicator 99 defines three categories of damage (end points): human health, ecosystem quality and depletion of resources. The quantification of inputs and outputs of systems is called Life Cycle Inventory (LCI). The LCIA converts these flows into simpler indicators.

The environmental impact rate B_j of the j-th material stream is calculated using its specific exergy ex_j, mass flow rate m_j and specific environmental impact b_k [15]:

$$B_j = E_j . b_j = m_j . e_j . b_j$$

(35)

Bj is expressed in Eco-indicator points per unit of time (Pts/s or mPts/s). The specific (exergy-based) environmental impact b_j is the average environmental impact associated with the production of the stream per unit of exergy of the same stream [Pts/(GJ exergy) or mPts /(GJ exergy)] [7]. Using the physical and chemical components of the specific exergy, the environmental impact rate B_j can be written as [15]:

$$B_j = m_j . e_{Ph} . b_{Ph} + m_j . e_{Ch} . b_{Ch} = m_j . b_j$$

(36)

The environmental impact rates associated with heat Q and work W streams are calculated as [15]:

$$B_Q = b_Q . E_Q$$

(37)

$$B_W = b_W . E_W$$

(38)

The exergy rate associated with heat transfer is calculated using the following equation [15]:

$$E_Q = \left(1 - \frac{T_0}{T_k}\right) . Q$$

(39)

where, T_0 is the ambient temperature and T_k the temperature at which the heat transfer crosses the boundary of the system. The objective of environmental impact balances is to calculate the environmental impact $B_{j,out}$ of all streams exiting each individual process. Thus, similar to an exergoeconomic analysis, the exergoenvironmental analysis is performed with a system of equations defined at the component level. The environmental impact balance for the k-th component of a power plant states that the sum of environmental impacts associated with all input streams plus the component-related environmental impact is equal to the sum of the environmental impact of all output streams [15]:

$$\Sigma_{j=1}^{n} B_{j,k}^{in} + Y_k = \Sigma_{j=1}^{m} B_{j,k}^{out}$$

(40)

The component-related environmental impact of the k-th component of the plant (Y_k) includes the three life-cycle phases of construction ($Y_{CO,k}$) (manufacturing, transport and installation), the operation and maintenance ($Y_{OM,k}$) and the disposal ($Y_{DI,k}$) [15]:

$$Y_k = Y_{CO,k} + Y_{OM,k} + Y_{DI,k}$$

(41)

Using data of the exergetic analysis and LCA, the specific environmental impact b_k is calculated as:

$$b_{k,in} = \frac{B_{k,in}}{E_{k,in}}$$

(42)

The equations used in the exergoenvironmental analysis of each component are shown in Table 4 [15].

Table 4: Environmental impact balances for the main components

Equipment	Environmental impact balance	Auxiliary equations
Compressor	$b_2.E_2 = b_1.E_1 + b_K.W_K + Y_K$ (43)	b_1 =0 (44) (fresh air)
Combustor	$b_3.E_3 = b_2.E_2 + b_{fuel}.E_{fuel} + (Y_{CC} + B_{CC}^{PF})$ (45)	b_{fuel} and b_{CO2}^{PF} [22]
Turbine	$b_4.E_4 + b_{TB}.W_{TB} = b_3.E_3 + Y_T$ (46)	$b_4 = b_3$(47) ; $b_K = b_{TB}$ (48)

The environmental impact balance applied to the k-th component of the power plant includes the environmental impact rates of product and fuel, $B_{P,k}$ and $B_{F,k}$. The environmental impact of exergy destruction in the power generation plant has been calculated by multiplying the exergy destruction with the specific environmental impact of the fuel of the plant. The environmental impact rate of fuel and product for the three components of the plant are shown in Table 5 [23].

Table 5: Environmental impact rate of fuel and product for the components of the power plant

Equipment	Environmental impact rate of fuel $B_{F,k}$ (mPts/s)	Environmental impact of product $B_{P.k}$(mPts/s)
Compressor	$b_K.W_K$	$b_2.E_2 - b_1.E_1$
Combustor	$b_2.E_2 + b_{fuel}.E_{fuel} + B_{CC}^{PF}$	$b_3.E_3$
Turbine	$b_3.E_3 - b_4.E_4$	$b_{TB}.W_{TB}$
OCGT	$b_{F,OCGT}.m_{fuel}.e_{fuel}$	$b_{P,OCGT} . W_{NET-OUT}$

The total environmental impact associated with component k includes the environmental impact of exergy destruction $B_{D,k}$ and the component-related environmental impact Y_k. In the case of the reactors, an additional term related to pollutant formation (PF) is added. Here, the environmental impact of pollutant formation (B_k^{PF}) is added to the combustor because it represents the account of pollutants formation such as CO, CO_2, CH_4, NO_x and SO_x [15].

$$B_{P,k} = B_{F,k} + Y_k + B_k^{PF}$$

(49)

Here, the pollutant formation is determined by the formed CO_2 emissions [15]:

$$B_{CC}^{PF} = b_{CO2}^{PF} . (m_{CO2,out} - m_{CO2,in})$$

(50)

The average specific (exergy-based) environmental impacts of product and fuel for the kth component are [15]:

$$b_{P,k} = \frac{B_{P,k}}{\hat{E}_{P,k}} \quad (51) \qquad b_{F,k} = \frac{B_{F,k}}{\hat{E}_{F,k}}$$

(52)

As commonly realized, the environmental impact of exergy destruction of each plant component is calculated multiplying it by the specific environmental impact of the fuel. Thus, the environmental impact rate of fuel of the k-th component ($B_{D,k}$) is defined as [15]:

$$B_{D,k} = b_{F,k} . E_{D,k}$$

(53)

The exergoenvironmental analysis does not only identify the components with the highest environmental impact, but it also reveals the possibilities and trends for improvement, in order to decrease the environmental impact of the overall plant. These trends can be identified using the relative environmental impact difference ($r_{b,k}$) and the exergoenvironmental factor ($f_{b,k}$) [15]. The environmental impact difference ($r_{b,k}$) of the k-th component of the power plant depends on the environmental impact of its exergy destruction ($B_{D,k}$) and its component-related environmental impact (Y_k) [15]:

$$r_{b,k} = \frac{1-EE}{EE} + \frac{Y_{k+} B_k^{PF}}{B_{D,k}} = \frac{b_{P.k} - b_{F,k}}{b_{F,k}}$$

(54)

$r_{b,k}$ is an indicator of the reduction potential of the environmental impact associated with the component. In general, a relatively high value of $r_{b,k}$ indicates that the environmental impact of the corresponding component can be reduced with a smaller effort than the environmental impact of a component with a lower value. Independently of the absolute value of the environmental impacts, the relative difference of specific environmental impacts represents the environmental quality of a component.

The sources of environmental impact formation in a component are compared using the exergoenvironmental factor $f_{b,k}$ that shows the relative contribution of the component-related environmental impact Y_k to the sum of its environmental impacts [15]:

$$f_{b,k} = \frac{Y_k}{Y_k + B_{D,k} + B_k^{PF}}$$

(55)

In the majority of the energy conversion systems, the value of $f_{b,k}$ has been shown to be negligible [24].

The environmental impact of electricity (EIE) of the Open Cycle Gas Turbine (OCGT) is estimated using the environmental impact balance applied to the overall system [15]:

$$EIE = \frac{3.6.(\,\mathrm{b_{F,tot}}\dot{\hat{E}}_{\mathrm{F,tot}}+\dot{\mathrm{Y}}_{\mathrm{tot}}+\dot{\mathrm{B}}_{\mathrm{tot}}^{\mathrm{PF}})}{\dot{\hat{E}}_{\mathrm{P,tot}}}$$

(56)

When the environmental impact associated with the exergy losses of the overall system is charged to the product, we obtain [15]:

$$EIE_T = \frac{3.6.(b_{F,tot}\dot{\hat{E}}_{F,tot}+\dot{Y}_{tot}+\dot{B}_{tot}^{PF}+B_{L,tot})}{\dot{\hat{E}}_{P,tot}}$$

(57)

PLANT DESCRIPTION

As shown in Figure 1, the power plant under investigation is an Open Cycle Gas Turbine divided into three different control volumes (Compressor, Combustor and Turbine). (1) Fresh air entering the compressor at point 1 is compressed to a higher pressure. (2) Upon leaving the compressor, air enters the combustion system at point 2, where fuel is injected and combustion occurs. The combustion process occurs essentially at constant pressure. (3) The combustion mixture leaves the combustion system and enters the turbine at point 3. (4) In the turbine section, the energy of the hot gases at point 3' is converted into shaft power in the shaft power generator.

Natural Gas Specifications

Natural gas is used in the power plant under investigation. The composition, heating values and chemical exergy of the natural gas at standard conditions are given in Table 6:

Table 6: Composition and properties of Natural gas

Component	Mass (%)	LHV (MJ/kg)	HHV (MJ/Kg)	e_{Ch} (MJ/kg)
CH_4	74	50.0	55.50	51.94
C_2H_6	13.1	47.80	51.90	50.0
C_3H_8	5.1	46.35	50.35	48.86
C_4H_{10}	3.2	45.75	49.50	48.27
N_2	4.1	0	0	0.05
CO_2	0.5	0	0	0.46
Total	100	47.1	52.02	48.93

Based on equation (4) and using the values of chemical exergies from Table 7, the chemical exergy of the natural gas at standard conditions is equal to 48.93 MJ/kg. The average values of the mass flow rate, temperature and pressure of the natural gas entering the combustion chamber are respectively 9.6 kg/s, 303 K and 2510 kPa. The value (-287.6 kJ/kg) of the thermo-mechanical energy of natural gas is evaluated using the following equation:

$$ex_{fuel}^{TM} = (h - h_0) - T_0\,(s - s_0)$$

(58)

From the given data and properties estimated using Hysys V8.6 with the Soave-Redlich-Kwong (SRK) equation of state , the specifications of the natural gas are given in Table 7

Table 7: Specifications of natural gas

T (K)	P(kPa)	Flow (kg/s)	h (kJ/kg)	s (kJ/ kg.K)	ex_T (kJ/kg)
303	2510	9.6	-4053.36	8.94	48642.5

AMBIENT AIR SPECIFICATIONS

The power plant is located in Abu Dhabi (UAE) and the exergy analysis was conducted during summer time. The average atmospheric conditions during summer in Abu Dhabi were: P= 100.8 kPa, T= 316 K and a relative humidity (RH) of 50%. Since fresh air is also the "dead state", its total exergy is equal to zero. The average mass flow rate m_{air} of air measured at the entrance of the air compression section is 484 kg/s. Based on the given information and properties estimated using Hysys V8.6 with the Soave-Redlich-Kwong (SRK) equation of state, the specifications of fresh air are shown in Table 8:

Table 8: Specifications of fresh air

T (K)	P(kPa)	Flow (kg/s)	h (kJ/kg)	s (kJ/ kg.K)	ex_T (kJ/kg)
316	100.8	484	-352.2	5.49	0

POWER PLANT EVALUATION

Exergy Analysis

The exergy analysis was conducted for summer conditions. The average summer atmospheric conditions are: P= 100.8 kPa, T= 316 K and relative humidity (RH)= 50% (absolute humidity 0.03 kg.m^{-3}). The operating conditions and the specific exergy of the streams (Figure 1) are shown in Table 9. The values of the exergy destruction and the exergy efficiency of each component of the power plant are obtained by solving the system of equations 8-15. The results are shown in Table 10.

Table 9: Stream-level results

Streams	T (K)	P (kPa)	Flow (kg/s)	h (kJ/kg)	s (kJ/ kg.K)	e_T (kJ/kg)
1 (ambient air)	316.0	100.8	484.0	-352.2	5.49	0 (dead state)
2 (compressed air)	714.2	1361.0	484.0	79.1	5.61	394.30
Fuel (natural gas)	303.0	2510.0	9.6	-4053.4	8.94	48642.50
3 (flue gas)	1480.0	493.6	493.6	-233.1	6.92	1133.80
4(exhaust gas)	886.00	100.80	493.6	-1003.00	7.04	337.44

Table 10: Component-level exergy results (summer weather conditions)

Equipment k	$\hat{E}_{D,k}$ (MW)	$\hat{E}_{D,k}$ (%)	$\hat{E}_{F,k}$ (MW)	$\hat{E}_{P,k}$ (MW)	$y_{D,k}$	EE_k (%)
Compressor	17.44	13.55	208.80	190.84	3.73	91.60
Combustor	98.15	76.28	657.81	559.64	21.0	85.10
Turbine	13.08	10.17	393.08	380.0	2.80	96.60
OCGT	**128.67**	**100.00**	**466.97**	**160.40**	**26.53**	**33.10**

The previous input data and obtained results are here verified through the simulation of the power plant using the software Aspen Hysys V8.6 with the Soave-Redlich-Kwong (SRK) equation of state. The simulation of the process is realized under the same operating and weather conditions (T=288 K, absolute humidity of 0.008 kg.m^{-3}). The results of the exergetic analysis and the values of the exergy of the fuel and product for each plant component are shown in Table 11. Figure 3 summarizes the effects of summer weather conditions on the exergy destruction ratio (y_D) of each component of the power plant.

Table 11: Component-level exergy results (design conditions)

Equipment k	$\hat{E}_{D,k}$ (MW)	$\hat{E}_{D,k}$ (%)	$\hat{E}_{F,k}$ (MW)	$\hat{E}_{P,k}$ (MW)	$y_{D,k}$	EE_k (%)
Compressor	14.94	11.90	183.80	166.35	3.20	92.00
Combustor	80.92	64.70	633.30	551.10	17.33	87.40
Turbine	29.12	23.30	391.92	362.8	6.24	92.60
OCGT	**125.0**	**100.00**	**466.97**	**168.6**	**26.77**	**34.60**

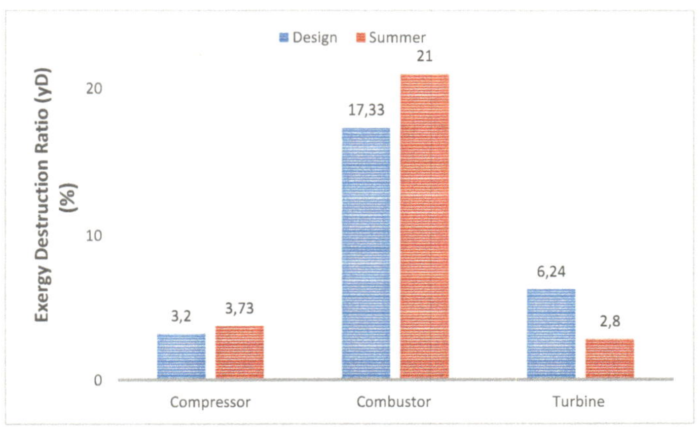

Figure 3: Effects of summer weather conditions on exergy destruction ratio (y_D)

Exergoeconomic Analysis

Using the operating conditions to solve equations (20) to (22), the capital cost rate of the three main components of the plant are respectively: Compressor (216 $/hr.), Combustor (25.4 $/hr.) and turbine (349.8 $/hr.). For a selected cost of natural gas of 2.6$/GJ [25] , an exergy of natural gas of 48.6 MJ/kg and the flow rate of 9.6 kg/s, the total cost of the fuel is estimated at 4232 $/hr. Solving the linear equations (23) to (28), the cost rate of all the streams can be calculated. The results of the analysis are shown in Table 12.

Table 12: Cost rate of the different equipment and streams of the plant

Stream	$\hat{C}$ ($/hr.)	$\hat{c}$ ($/GJ)
W_K	3120.3	4.15
W_T	5678.7	4.15
1	0.0	0
2	3336.3	4.85
Fuel	4232	2.6
3	7593.7	3.77
4	2264.8	3.77

Finally, equations (29) to (34) are utilized to estimate the exergoeconomic parameters of the different equipment of the gas turbine during summer conditions and for design conditions (Tables 13-14). Figure 4 summarizes the effects of summer weather conditions on the relative cost difference (rk) of each equipment of the power plant.

Table 13: Exergoeconomic parameters for each equipment (summer weather conditions)

Equipment	EE (%)	ED (MW)	Y_D (%)	$\hat{Z}_K$ $/hr.	$c_{F,k}$ $/GJ	$c_{P,k}$ $/GJ	$\hat{C}_{D,k}$ ($/hr.)	$\hat{C}_{D,k} + \hat{Z}_k$ ($/hr.)	r_k (%)	f_k (%)
Compressor	0.916	17.96	3.85	216.0	4.15	4.85	260.6	476.6	16.9	45.3
Combustor	0.851	98.17	21.04	25.4	3.19	3.77	1127.2	1152.6	18.2	2.2
Turbine	0.938	13.08	2.80	349.8	3.76	4.15	330.5	680.3	10.4	51.4

Table 14: Exergoeconomic parameters for each equipment (design conditions)

Equipment	EE (%)	ED (MW)	Y_D (%)	$\hat{Z}_K$ $/hr.	$c_{F,k}$ $/GJ	$c_{P,k}$ $/GJ	$\hat{C}_{D,k}$ ($/hr.)	$\hat{C}_{D,k} + \hat{Z}_k$ ($/hr.)	r_k (%)	f_k (%)
Compressor	0.916	15.45	3.74	216.0	4.24	4.85	235.83	451.82	14.8	47.8
Combustor	0.872	82.2	17.61	25.4	3.18	3.67	941.02	966.42	15.4	2.70
Turbine	0.896	29.12	6.24	349.8	3.68	4.24	385.78	735.58	15.2	47.54

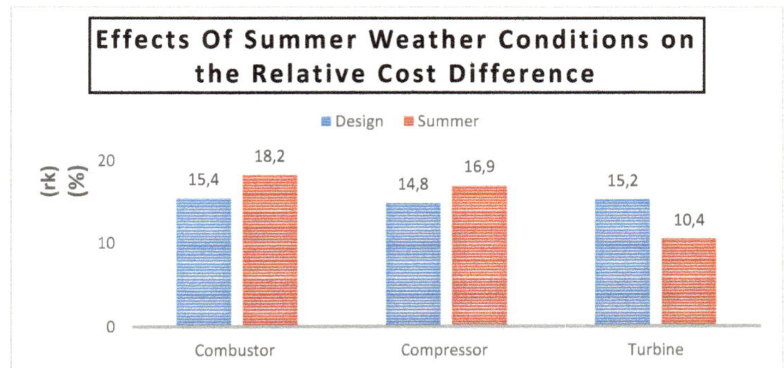

Figure 4: Effects of summer weather conditions on the relative cost difference (r_k)

Exergoenvironmental analysis

The specific environmental impact of carbon dioxide and the depletion of fuel in Eco-99 points were selected from literature [23]. As shown in Figure 2, three end-point categories of the LCIA are considered: Damage to human health, damage to ecosystem quality and damage to fossil resources.

Global warming (kg(CO2-eq.)/kWh): This indicator measures the total quantity of greenhouse gases (GHG) released to the atmosphere from the power plant. The value of the specific environmental impact of CO_2 for Eco-99 is equal to 5.454 mPts/kg [23].

Depletion of fossil fuel: This indicator measures the total primary energy in fossil resources used for the production. When no pollutants are considered, the value of 3.5 mPts/MJ can be used. In order to take into account formed pollutants, the value of b_{fuel} equal to 5.38 mPts/MJ is used. This value includes the environmental impact of pollutant formation [23].

It has been shown that the component-related environmental impact (Y_k) is negligible in an exergoenvironmental analysis [15, 23]. Thus, it has not been considered here. Based on collected data and specified assumptions, the values of the environmental impact rate Bj and the specific (exergy-based) environmental impact bj of all the streams are obtained by solving the system of Equations (43)-(48). The results are shown in Table 15. Equations (49)-(54) are used to estimate the exergoenvironmental parameters of the different components of the OCGT both summer conditions and design conditions (Tables 16 and 17). Figure 5 summarizes the effects of summer weather conditions on the environmental impact difference (r_b) of each equipment of the power plant.

Table 15: Stream-level environmental impact rate

	Summer conditions		Standard conditions	
Stream	b_j (mPts/MJ)	B_j (mPts/s)	b_j (mPts/MJ)	B_j (mPts/s)
W_K	5.32	1110.60	5.42	992.50
W_T	5.32	2021.60	5.42	1959.10
1	0	0	0	0
2	5.82	1110.60	5.97	993.10
Fuel	3.50	1634.40	3.50	1634.40
3	5.14	2879.00	5.01	2761.00
4	5.14	856.10	5.01	797.50

Table 16: Exergoenvironmental parameters (summer weather conditions)

Equipment	$b_{F,k}$ (mPts/MJ)	$b_{P,k}$ (mPts/MJ)	$B_{D,k}$ (mPts/s)	B_L (mPts/s)	B_k^{PF} (mPts/s)	$B_{D,k} + B_k^{PF} + B_L$ (mPts/s)	$r_{b,k}$ (%)
Compressor	5.32	5.82	92.78	0	0	92.78	9.4
Combustor	4.38	5.14	429.90	0	133.97	563.27	17.4
Turbine	5.14	5.32	67.36	0	0	67.36	3.5
OCGT	**3.50**	**16.40**	**450.34**	**856.10**	**133.97**	**1440.40**	**367.0**

Table 17: Exergoenvironmental parameters (standard conditions)

Equipment	$b_{F,k}$ (mPts/MJ)	$b_{P,k}$ (mPts/MJ)	$B_{D,k}$ (mPts/s)	B_L (mPts/s)	B_k^{PF} (mPts/s)	$B_{D,k} + B_k^{PF} + B_L$ (mPts/s)	$r_{b,k}$ (%)
Compressor	5.42	5.97	80.97	0	0	80.97	10.1
Combustor	4.37	5.00	353.60	0	133.97	487.60	14.4
Turbine	5.00	5.42	145.60	0	0	145.60	8.4
OCGT	**3.50**	**15.20**	**437.50**	**818.00**	**133.97**	**1389.50**	**335.00**

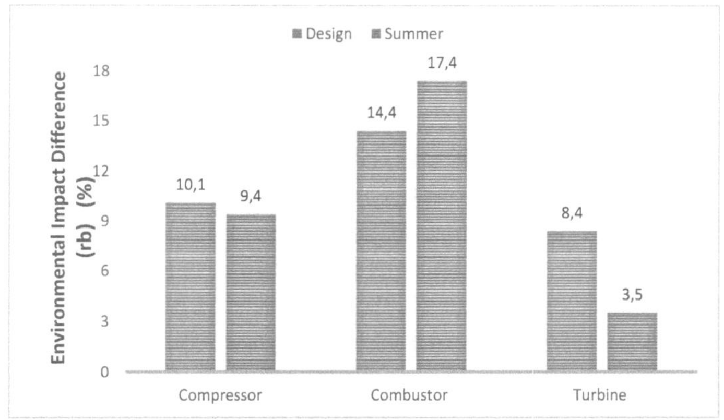

Figure 5: Effects of summer weather conditions on Environmental Impact Difference (r_b)

Based on equations (56) and (57), the environmental impact of a kWh of electricity during summer weather conditions is 40.3 mPts/kWh (exergy destruction only) and 59.0 mPts/kWh (including the exergy loss). The corresponding values related to the standard weather conditions are 37.8 mPts/kWh and 54.7 mPts/kWh, respectively. Lastly, summer weather conditions increase the total environmental impact of the power plant by 6.6% (without exergy loss) and 7.9% (including exergy loss). The effects of summer conditions on the environmental impact of electricity produced by the power plant are shown in Figure 6.

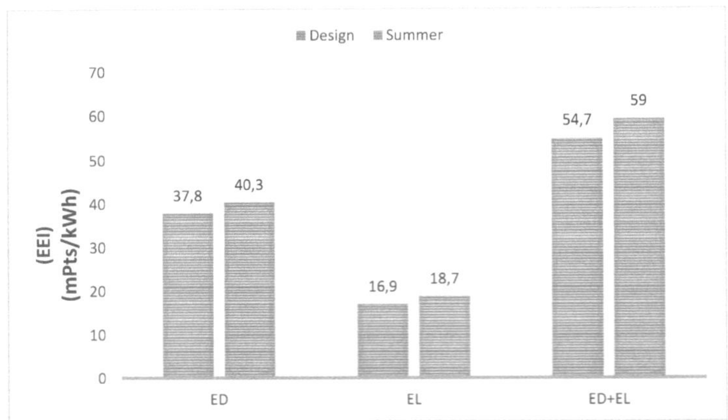

Figure 6: Effects of summer weather conditions of environmental impact of electricity produced (ED: exergy destruction, EL: exergy loss)

VALIDATION OF THE SIMULATION RESULTS

According to results obtained using Hysys V8.6, the power generated by the turbine and the power needed by the compressor for summer conditions are 380 MW and 208.8 MW, respectively. Assuming a mechanical efficiency for the turbine and compressor equal to 98% and a generator efficiency equal to 98%, the power plant generates 160.4 MW net under summer conditions. The net output of the operating power plant under standard conditions is 165 MW while our simulation indicates that the net power produced at design conditions is 168.6 MW. The relative shift of 2.2% presents an acceptable margin of error.

ANALYSIS OF RESULTS

Exergy Analysis

According to results shown in Tables 10 and 11, (1) the combustor is the main contributor to the exergy destruction of the power plant (Y_D =17.33%) and has therefore the lowest exergy efficiency (EE= 87.4%). Summer weather conditions increase its exergy destruction ratio by 21.2 % and decrease its exergetic efficiency by 2.6%. (2) The turbine has the second highest exergy destruction ratio (Y_D =6.24%). Due to important positive effects of higher temperatures at the entrance of the turbine, the summer weather conditions decrease its exergy destruction ratio by 55% and increase its exergetic efficiency by 4.3%. (3) The compressor has the lowest contribution to the exergy destruction of the power plant (Y_D =3.2%). Unlike the turbine, the summer weather conditions increase the exergy destruction ratio of the compressor by 16.5% and decrease its exergetic efficiency by only 0.4%. (4) For the overall OCGT, summer weather conditions decrease its exergy destruction ratio by 0.9%, from 26.7% to 26.5% but decrease its exergetic efficiency by 4.3%, from 34.6% to 33.1% and its net power output by 4.9%, from 168.6 MW to 160.4 MW.

Exergoeconomic Analysis

In Tables 13-14 indicate that (1) the combustor has the highest contribution to the cost of the final product (r=14.8) and the lowest contribution of the capital investment (f_k= 2.70). Summer weather conditions increase the cost rate of exergy destruction ($\widehat{C_D}$) by 19.8%. and the relative cost difference (r_k) has increased by 18.2 %. The summer atmospheric conditions have decreased its Exergoeconomic factor (f_k) by 18.5%. The analysis of these variables indicate that the high cost of the product is mainly due to the cost of exergy destruction within the combustor. This situation becomes more important during summer conditions. (2) The turbine has the second highest contribution to the cost of the final product (r=15.2) and the second highest contribution of the capital investment (f_k= 47.5). The summer weather conditions have decreased the cost of exergy destruction of the turbine by 14.3 %. and its contribution to the cost of the final product (r) by 31.6%. The exergoeconomic factor has increased only by 8.1%. Since the turbine has the highest capital cost rate, the large decrease of its exergy destruction (55%) during summer conditions did not have an important impact on the value of the exergoeconomic factor. (3) The compressor has the lowest contribution to the cost of the final product (r=14.8) and the highest contribution of the capital investment (f_k= 47.8 %). The summer conditions increased the cost rate of exergy destruction and the relative cost factor respectively by 10.5% and 14.2%. On the other hand, the exergoeconomic factor has decreased by 5.2%. These findings are in concordance with the increase of the exergy destruction of 16.5% during summer conditions.

Exergoenvironmental Analysis

In agreement to the exergetic analysis, the results of the exergoenviromental analysis (Tables 16 and 17) indicate that (1) the combustor presents the highest environmental impact of exergy destruction (B_D= 353.6 mPts/s). The summer weather conditions further increase this impact by 21.6%. In addition, the combustor also has the highest contribution to the total environmental impact of the final product (rb =14.4%), while summer weather conditions increase this contribution by 20.8%. (2) The compressor has the lowest environmental impact of exergy destruction (B_D=80.97 mPts/s) and summer weather conditions increase this impact by 14.6%. The compressor has the second highest contribution to the total environmental impact of the final product (rb =10.1%). Unlike the combustor, the data indicate that summer weather conditions decrease this contribution by 7.4%. (3) The expander has the second highest environmental impact of exergy destruction and summer weather conditions decrease this impact by 53.7%. The expander has the lowest contribution to the total environmental impact of the final product (rb =8.4%), while summer weather conditions decrease this contribution by 58.3%. (4) The environmental impact of a kWh of electricity of the power plant under standard conditions was 37.8 mPts/kWh (exergy destruction only), and 54.7 mPts/kWh (both exergy destruction and exergy loss). Summer weather conditions increased these impacts by 6.6% (exergy destruction only) and 7.9% (for both exergy destruction and exergy loss).

RECOMMENDATIONS

First, the recoverable performance loss in the plant can be easily rectified by water washing or, more thoroughly, by mechanically cleaning the compressor & turbine blades and vanes. Second, the non-recoverable loss of performance caused by reduction in component efficiencies, could be corrected by replacement of affected parts during inspection intervals.

Based on the results obtained by the exergy-based analysis of the power plant under design conditions, the company should focus on the combustion process because the combustor is the main source to the exergy destruction of the power plant. It has also the highest contribution to the cost of the final product and the highest contribution to the environmental impact of the power plant. In order to reduce the effects of irreversibilities in the combustion process on the performance of the plant, it is suggested to: (1) check if the fuel composition meets the original equipment manufacturer (OEM) specification because a number of factors including auto-ignition, flame temperature, emissions and stability depend on fuel specifications. (2) replace the feedforward control (Ratio control) by a feedback control with continuous measurement of both O_2 and CO leaving the combustor. This new process control strategy could provide the needed information for effective combustion for significant energy savings & minimizing excess air.

As expected, the comparison between the results obtained under the design conditions and during summer weather conditions indicate that the overall performance of the power plant is lower during the summer season and its environmental impact is higher. It was found that the negative effects of summer weather conditions on the performance of the combustion chamber and the compressor are partly compensated by their positive effects on the performance of the turbine. It is suggested to add a cooling system at the entrance of the compressor to bring atmospheric air conditions during summer season closer to the standard conditions. Finally, in order to take advantage of the high temperature of the exhaust gases (886 K), the addition of a heat recovery steam generator (HRSG) to generate steam is recommended. The generated steam can be further used in a Ranking cycle to produce additional electricity. The final goal is to increase the output of the plant and decrease its total environmental impact through the increased output.

REFERENCES

1. Todorova V. UAE released 200 million tonnes of greenhouse gases in 2013. The National (UAE), January 21, 2015.
2. Aljundi, I.H.: Energy and exergy analysis of a steam power plant in Jordan. Applied Thermal Engineering 29: 324–328 (2009)
3. Meyer, L., Tsatsaronis, G., Bushgeister, J. and Schebek, L. (2009) Exergoenvironmental analysis for evaluation of the environmental impact of energy conversion systems, Energy - The International Journal 34, 75-89.
4. Petrakopoulou F., Tsatsaronis G., Boyano A. and Morosuk T. (2010b) Exergoeconomic and Exergoenvironmental Evaluation of power plants including CO2 capture, Carbon Capture & Storage Special Issue of Chemical Engineering Research and Design, In press, DOI: 10.1016/j.cherd.2010.08.001.
5. Keenan, J. H. (1932) A steam chart for second law analysis, Trans. ASME, 54 (3), 195–204.
6. Tsatsaronis, G. (1984) Combination of Exergetic and Economic Analysis in Energy Conversion Processes, in: Energy Economics and Management in Industry (Reis, A., Smith, I., Peube, J. L., Stephan, K., Eds.), Pergamon Press, Oxford, 151-157.
7. Bejan A; Tsatsaronis G., Moran M: Thermal Design and Optimization; J. Wiley & Sons Edition. (1996)
8. Moran, M.J. , Shapiro H.N.: Fundamentals of Engineering Thermodynamics, John Wiley & Sons, 6th edition (SI Units), (2010)
9. Rahman, M.M., Ibrahim T.K., Abdalla A. N. : Thermodynamic performance analysis of gas-turbine power-plant. International Journal of the Physical Sciences Vol. 6(14), pp. 3539-3550, 18 July, 2011

10. Cihan A , Hacıhafızoglu O, Kahveci K: Energy–exergy analysis and modernization suggestions for a combinedcycle power plant, International Journal of Energy Research 30 (2) PP: 115–126 (2006)

11. Roosen P. , Uhlenbruck S, Lucas K : Pareto optimization of a combined cycle power system as a decision support tool for trading investment vs. operating costs; International, Journal of Thermal Sciences 42 (6) PP: 553–560 (2003).

12. Moran J. : Availability Analysis: A Guide to Efficient Energy Use, Prentice-Hall, pp. 199–215 (1982).

13. Lehtinen H, Saarentaus A, Rouhiainen J, Pitts M, Azapagic A,. A Review of LCA Methods and Tools and their Suitability for SMEs; Eco-Innovation. BIOCHEM, May 2011.

14. Sciubba E. Beyond thermoeconomics: The concept of extended exergy accounting and its application to the analysis and design of thermal systems. Int. J. Exergy 2001; 2: 68–84.

15. Meyer L, Castillo R, Buchgeister J ,Tsatsaronis G. Application of Exergoeconomic and Exergoenvironmental Analysis to an SOFC System with an Allothermal Biomass Gasifier. Int. J. of Thermodynamics 2009; 12 (No. 4):177-186.

16. Buchgeister J, Exergoenvironmental Analysis - A new approach to support design for environment of chemical processes. Chemical Engineering & Technology 2010; 33, No. 4: 593-602, 2010.

17. Rimos S, A,Hoadley, Brennan D. Environmental consequence analysis for resource depletion. Process Safety and Environmental Protection Journal 2014; 92 (6): 849–86, 2014.

18. Goedkoo M, Spriensma R. The Eco-indicator99: a damage oriented method for life cycle impact assessment. Methodology report, Amersfoort, Netherlands, 2000.

19. Guinee J.B. Editor, Life cycle assessment: an operational guide to the ISO standards; LCA in perspective; guide; operational annex to guide. The Netherlands: Centre for Environmental Science, Leiden University; 2001.

20. Udo de Haes H.A . Life cycle impact assessment—striving towards best practice. Pensacola: Society of Environmental Toxicologyand Chemistry (SETAC); 2002.

21. Jolliet O. Life cycle impact assessment definition. Study of the SETAC-UNEP life cycle initiative.UNEP,2003.

22. Petrakopoulou, F. Comparative Evaluation of Power Plants with 753 CO2 Capture: Thermodynamic, Economic and Environmental Performance. Ph.D. Thesis, Technische Universität Berlin, 2010; p 230.

23. Morosuk T, Tsatsaronis G, Koroneos C, On the Effect of Eco-indicator Selection on the Conclusions Obtained from an Exergoenvironmental Analysis. Proceedings of ecos 2012 - the 25th international conference on efficiency, cost, optimization, simulation and environmental impact of energy systems; June 26-29, 2012, Perugia, Italy.

24. Szargut J. Minimization of the consumption of natural resources. Bull Pol Acad Sci: Tech Sci. 1978; 26(6):41–46.

25. Amin A. Z. and Al Zeyoudi, T: Renewable energy prospects for the United Arab Emirates, Masdar Institute (April, 2015)